CRAZY LAZY CORN!!!

A Book about Plant Genetics for Students & Future Scientists

AF718663

Crazy Lazy Corn!!!

A Book about Plant Genetics for Students & Future Scientists

Copyright ©2019

Hank Bass & Kome Onokpise
Florida State University
Florida Agricultural and Mechanical University

ISBN: 978-09793451-9-7

Library of Congress Cataloging-in-Publication Data
Hank Bass and Kome Onokpise

Authored by Hank Bass, Kome Onokpise & Florida A&M University and Florida State University. All rights reserved. Unless otherwise noted, no part of this book may be reproduced, stored in a retrieval system, transmitted in any form or by any means, electronic, mechanical, photocopy, or recording without express written permission from the authors, except for brief quotations or critical reviews.

SokheChapke Publishing

TABLE OF CONTENTS

TABLE OF CONTENTS (cont'd)

Preface

The goal of this book is to share the wonder of science and plant genetics with young students, teachers, and popular audiences. We hope it will stimulate curiosity about plants, using corn (maize) mutants as a means to achieve this goal. Maize and corn are the same thing. Plant scientists, such as the authors of this book, often use the word "maize," but the general public more frequently use the word "corn." Both maize and corn are correct as common names. The scientific name for the genus and species is *Zea mays*. For nearly 20 years, the authors and university professors Drs. Bass and Onokpise have worked with middle and high school kids during the summer months demonstrating plant genetics and breeding practices. Because the students showed great curiosity and enjoyment working with and learning about maize mutants in the field and laboratory, we decided to make this book. It seeks to engage the reader with fun facts and links to contemporary research topics and resources. This unique book reflects our vision of sharing the love of maize mutants that has inspired countless students, scientists, and farmers for over 100 years. An equally important goal of this book is to explain how knowledge of plant genetics is key to solving many of the grand challenges for future generations, such as world hunger.

The content ranges widely from the origins of agriculture to Mendelian inheritance and its role in the Green Revolution and McClintock's discovery of jumping genes. This book also teaches about DNA, genetics, and mutations. We highlight some of the students' favorite mutants, such as the lazy, knotted, dwarf, albino, and wrinkled sweet corn mutants. The book also touches on the importance of maize in culture and art, and ends with information on how to obtain and grow your own maize mutants. Throughout the book, we have included links to additional information related to these topics, which we hope will be useful for teaching in the science, technology, engineering, and math (STEM) disciplines while addressing next-generation science standards.

This book, made possible by many people, was initially inspired by the beautiful monograph "Mutants of Maize" by Neuffer, Coe, and Wessler (Cold Spring Harbor Press, 1996). We are particularly grateful to Dreamal Worthen, Kimberly Davis, Karen McGinnis, Jonathon Doster, Kristin Bussey, Gregg Hoffman and Cheryl Jennings for their invaluable contributions to the organization, content, and design of the book and supporting the vision. We also thank the Forestry and Conservation Education (FACE) program students and the National Science Foundation for their enthusiasm and participation in the mutant field day events. A glossary of many of the most important or technical terms is included to enhance the instructional and teaching value of this book. Terms in the glossary are bolded when they first appear.

Acknowledgement

This work was supported by the Department of Biological Science at Florida State University and the College of Agriculture and Food Sciences at Florida A&M University. We are indebted to the many guidance counselors, FACE counselors, and science teachers from area high schools in Leon, Jefferson, Wakulla, and Gadsden counties, who have encouraged their students to participate in the FACE program for 20 years. We thank Marty Sachs and Philip Stinard at the Maize Genetics Cooperation - Stock Center (USDA/ARS, at the University of Illinois, Urbana/Champaign) for providing mutant seed stocks. We are also very thankful for the advice, moral, and financial support for the outreach activities funded by the National Science Foundation Plant Genome Research Program (Awards DBI-0321639, IOS-1025954, and IOS -1444532), the USDA 1890 Capacity Building Grants Program (USDA-CSREES-AFRI Award 98-38820-6385), and USDA-Forest Service-SRS (Awards SRS-06DG1133012599, SRS-14DG113301079), for their contributions to FACE program. We thank Jen Kennedy from the Department of Biological Science at The Florida State University for editorial assistance. We thank Jen Kennedy from the Department of Biological Science at The Florida State University for editorial assistance. Finally, we thank Dr. Cheryl Jennings and our families for their support and encouragement, which made this project possible and fun.

I. HISTORY AND CHALLENGE OF AGRICULTURE

Introduction

This section introduces the reader to the historical origins of plant agriculture, the importance of "The Green Revolution," which was led by a plant scientist named Norman Borlaug, and challenges for future plant scientists. The last section emphasizes the challenges we face in continuing to feed the world as the population is expected to reach 10 billion by the year 2050.

Chapters in this section include:

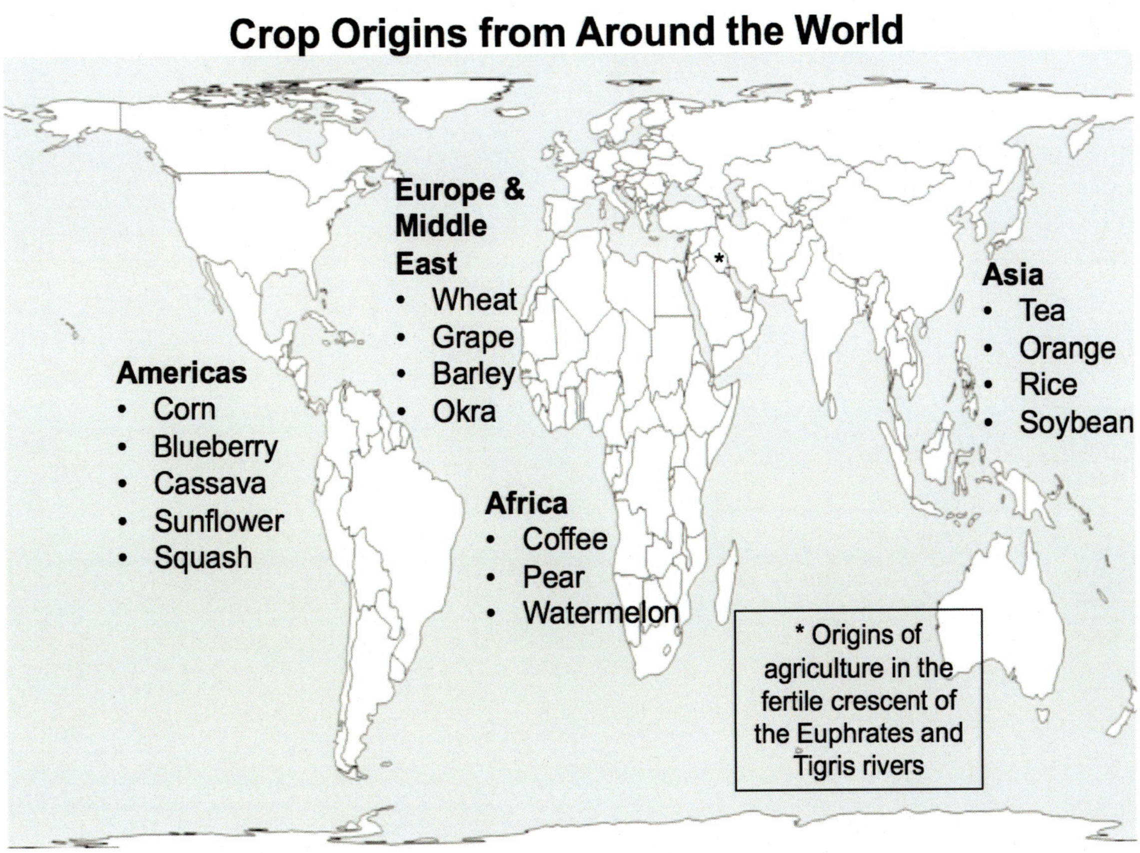

Section I: Figure 1. World map showing the origins of domesticated crops. Crops originally from specific world regions are now grown all over the world. The background map is a Gall-Peters equal area projection map (from http://metrocosm.com/mercator/).

Chapter 1 The History of World Agriculture

DO YOU KNOW WHERE IN THE WORLD AGRICULTURE STARTED?

The answer is the *Fertile Crescent*. The *Fertile Crescent* was a region in the Middle East and is actually shaped like a crescent. History tells us that agriculture originated there, in an area once known as Mesopotamia, near important rivers in the Middle East and Egypt. In Mesopotamia, people known as the Sumerians developed elaborate irrigation systems that made it possible for them to grow many crops. The two rivers in this region are named the *Tigris* and the *Euphrates*. Additionally, the *Fertile Crescent* was very close to the *Mediterranean Sea*, the *Red Sea*, and the *Persian Gulf*.

Can you find where the Fertile Crescent was once located in the Middle East?

The origins of modern agricultural practices trace back to a region of the world called "the fertile crescent," a region located in the Middle East.

Some of our most important foods come from this region, including wheat, barley, rye, oats, lentils, fava beans, and peas. To learn more about the history of this region and the origins of agriculture, visit https://www.icarda.org/fertile-crescent

Crop domestication: Similarly, agricultural practices were occurring in other regions of the world. For instance, rice domestication occurred in the Yangtze river valley in China; squash and maize cultivation occurred in Mexico and Central America; coffee and watermelon in Africa; and sugar beet in Europe (Section I: Figure 2 and Section I: Figure 3). These activities occurred a long time ago, more than 10,000 years ago. For more information on the domestication of maize, check out this excellent video:

http://media.hhmi.org/biointeractive/films/PoppedSecret.html

Section I: Figure 2. A Great Story of Plant Domestication. Comparison of the wild ancestor of maize (Teosinte) and the modern domesticated maize (Modern Corn). Image by Nicolle Rager Fuller.

In addition to specific plants, agriculture also includes practices, such as row cropping, ground preparation, irrigation, harvest, seed collection and re-use, and breeding and selection.

Early **agricultural mechanization** involved simple tools such as plows, and the use of animals such as oxen to help plow and carry out other farm operations.

Later, the industrial revolution allowed for a transition to machine-based agriculture, using tractors for planting, plowing, and harvesting (Section I: Figure 4).

Fun Fact: Did you know that the early ads for tractors said:

"***Better and cheaper than horses***"

The Detroit Tractor Company was known to advertise

"It drives like a horse; does the work of 10 horses, but eats only when at work"

For more information: https://en.wikipedia.org/wiki/Mechanised_agriculture

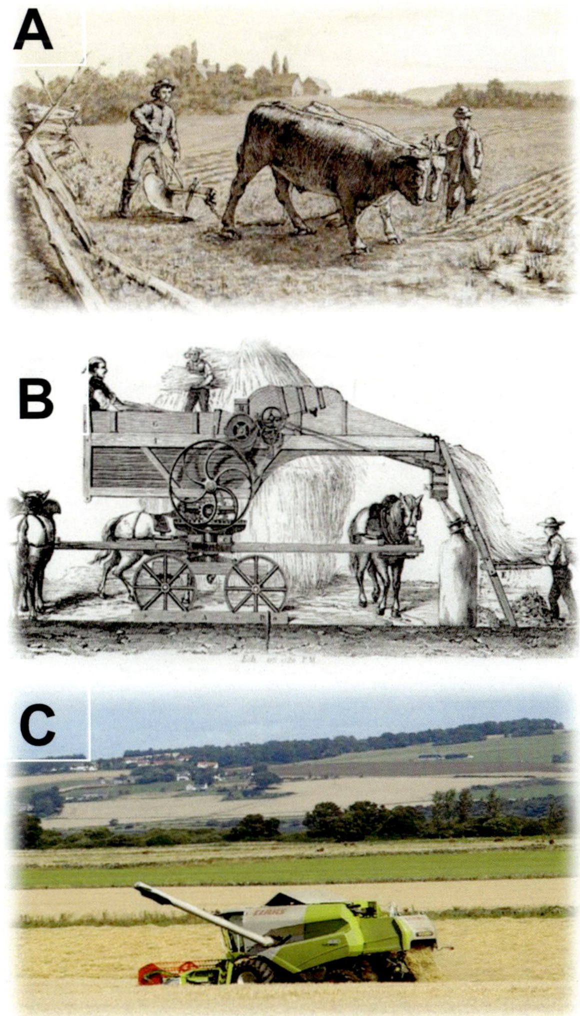

Section I: Figure 3. Agricultural Mechanization: from farm animals to machines.

A. Farming animals; early plowing with oxen (early 1800s) https://www.flickr.com/photos/lac-bac/31367043552/in/album-72157677492475586/ Image may be freely reproduced and modified.

B. Threshing machine used to improve wheat harvest (late 1800s) Image may be freely reproduced and modified. https://commons.wikimedia.org/wiki/File:Batteuse_1881.jpg

C. Harvesting wheat with a modern combine harvester machine (20th Century) Combine harvester By Brian Forbes Commons.wikimedia.org/wiki/File:Claas_Tucano_430.jpg Image may be freely reproduced and modified.

Chapter 2 The Green Revolution, Led by Nobel Laureate Norman Borlaug

HAVE YOU EVER HEARD OF THE GREEN REVOLUTION?

The Green Revolution refers to research and technology in plants and agriculture that greatly increased food production for a 30-year period starting in the 1930s.

DID YOU KNOW WHO IS "THE FATHER" OF THE GREEN REVOLUTION?

Norman Ernest Borlaug (1914 –2009), an American agronomist and humanitarian who saved people all over the world from starvation.

The man who saved a billion lives.

Section I: Figure 4. Norman Ernest Borlaug (born March 25th, 1914). He was awarded the Nobel Peace Prize in 1970 for his contributions to the Green Revolution. From Wikimedia Commons. http://www.fas.usda.gov/icd/stconf/photos/enlargements/borlaugnl062403.jpg; David Levy; http://www.fas.usda.gov/icd/stconf/photos/enlargements/borlaugn1062403.jpg{{PD-USGov}}. Image may be freely reproduced and modified.

His great achievements were made possible by what he learned as a college student at the University of Minnesota. He received his first college degree in 1937, a Bachelor of Science (B.Sc.) in Forestry. He received his next college degree in 1942, a Doctor of Philosophy (Ph.D.) in plant pathology and genetics. This education prepared him for his revolutionary work in agricultural research. He was awarded the **Nobel Peace Prize** in 1970 for his contributions to world peace through increasing food supply, the first time a plant scientist received the Nobel Peace Prize.

HOW DID DR. BORLAUG INCREASE THE WORLD FOOD SUPPLY?

He and others developed **semi-dwarf**, high-yield, disease-resistant wheat varieties while he was working in Mexico as an agricultural researcher. The semi-dwarf trait is genetic, caused by a heritable DNA mutation.

Section I: Figure 5 . Wheat plots in a cereal breeding trial – tall and short varieties.
Picture courtesy of Nigel Cattlin/Alamy Stock Photo.

Productive varieties were combined with advances in agricultural mechanization (Chapter 2) in Mexico, India, and Pakistan. As a result, Mexico became a major exporter of wheat, while Pakistan and India doubled their wheat production.

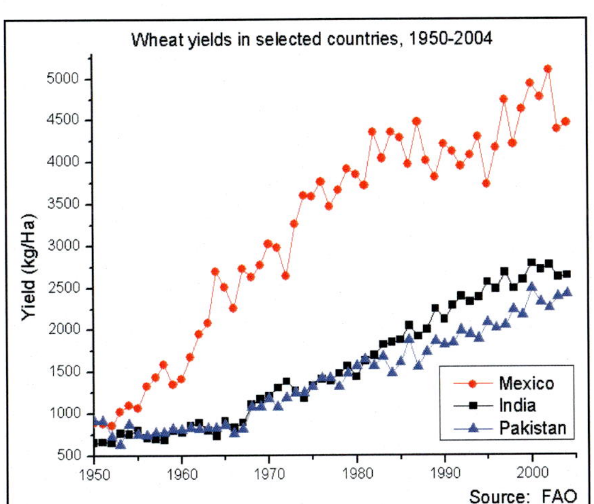

Section I: Figure 6. Wheat yield increases over time. Plot showing the increase in wheat yields in Mexico, India, and Pakistan from 1950 to 2004. The Y axis shows yield in kg wheat per hectare. The X axis shows time in years. Image courtesy of Wikimedia commons at https://commons.wikimedia.org/w/index.php?curid=5152133.

This story illustrates how a plant scientist can work with *research science teams* (Chapter 3) to achieve great things for humanity. They made so much progress improving and securing food around the world that Dr. Borlaug's Green Revolution is credited with saving over a billion people from starvation.

CAN YOU HELP US WITH THE NEXT GREEN REVOLUTION?

IF SO, WHAT MIGHT YOU DO?

THE WORLD NEEDS YOUR HELP.

For more information and inspiration, please visit the October 2014 National Geographic Article on "**the next green revolution**" to learn concepts and definitions of plant breeding and new bio- technologies.

http://www.nationalgeographic.com/foodfeatures/green-revolution/

Chapter 3 Solving the Future Food Crisis through Science & Technology

At the time of the Green Revolution, the world population was around 3 billion people. In the next 30 years, the world population is projected to reach nearly 10 billion. This creates a major challenge for humanity and our future.

HOW WILL WE FEED EVERYONE?

Section I: Figure 7. Courtesy of https://quantumleaders.com

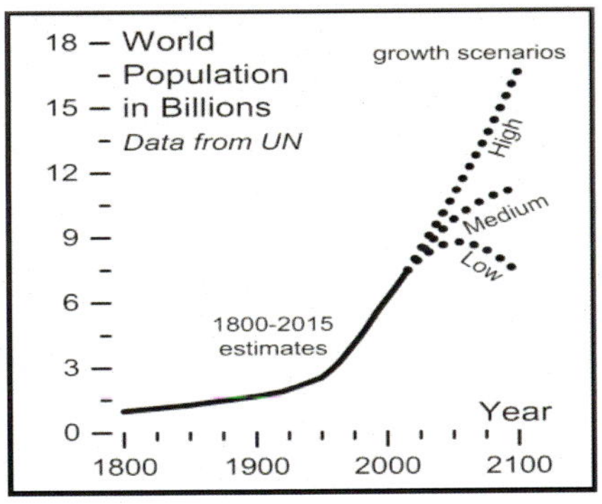

Section I: Figure 8. World population estimates from 1800 to 2100, based on "high", "medium" and "low" United Nations projections in 2015 and UN historical estimates for pre-1950 data. (Image from By Bdm25 - Own work, CC BY-SA 4.0, from https://commons.wikimedia.org/w/index.php?curid=51036438). This image may be freely reproduced and/or modified.

Just as plant breeding innovations helped us deal with hunger and famine in the 20th century, teams of people are needed to work together globally with each other and with agencies, universities, and companies to meet these challenges in the future.

Which type of team would you choose?

Biotechnologists – we produce & improve crop varieties for the future, bioengineered for improved traits

Botanists – we study plants & share knowledge

Entomologists – we discover plant-insect interactions

Farmers – we produce, grow, & harvest food for animals and people

Food processing & distribution – we prepare crops for shipping food to market

Food Scientists – we process & evaluate food crops for consumption

Geneticists & Breeders – we discover how traits are shuffled & passed through families

International Agencies – we coordinate global plant and animal products for people in different countries

Mathematicians – we model plant processes and discover biological rules

Pathologists - we discover how plants fight off diseases

Physiologists – we discover how plants grow and develop

Statisticians – we analyze numerical data for meaningful patterns

Section I: Figure 9. Working on maize genetics. Drs. Bass and Onokpise study maize with help from the National Science Foundation's Plant Genome Research Program.

Section I: Figure 10. Plant Biotechnology. A rose plant shown here is grown from cells in tissue culture. Photo courtesy of Wikimedia commons: https://commons.wikimedia.org/w/index.php?curid=2970936.

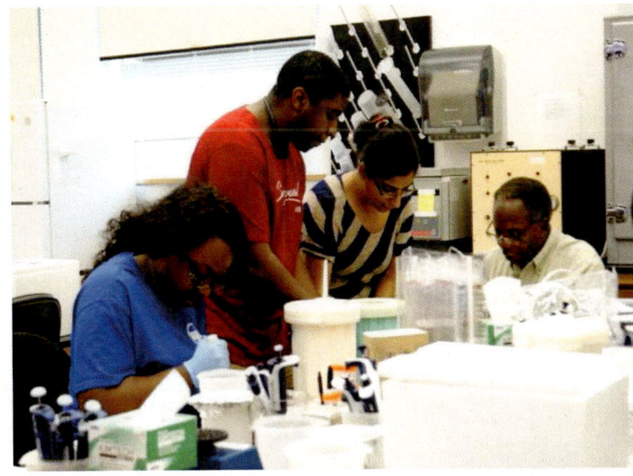

Section I: Figure 11. Research teams working on maize genetics and genomics.
The NUPRIME (Nuclease Profiling in Maize Epigenomics) Workshop, hosted by FSU and FAMU, trained scientists and students from around the country. Here people ranging from college students to university professors were learning how to isolate and study the genetic material of maize.

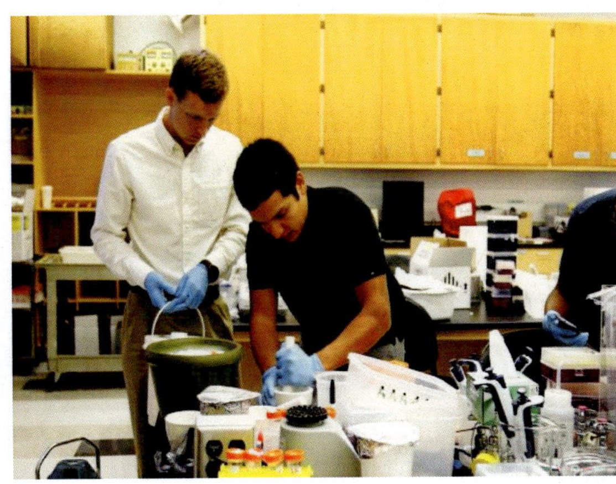

DID YOU KNOW THAT TEAMS OF PEOPLE MET TO PRODUCE A 10-YEAR VISION TO GUIDE PLANT SCIENCE RESEARCH FROM 2015-2025?

The Plant Science Summit was planned to "engage the broad plant science research community in developing a consensus plan to invigorate and guide plant science research over the next decade". To learn more about the Summit please see the report called "The Plant Science Decadal Vision" at: https://plantsummit.files.wordpress.com/2013/07/plantsciencedecadalvision10-18-13.pdf.

Key points from this report include:

- ***Population growth comes with need for increased food production.***
- ***There will be a lot of jobs and career opportunities if you want to join these plant science teams.***
- ***Storing germplasm is very important.***

Germplasm refers to resources such as seed collections and seed banks and tree nurseries. A germplasm collection iscrucial for the maintenance of **biological diversity.** For more information, visit https://en.wikipedia.org/wiki/Germplasm

II. PLANT SCIENCE

Introduction

This section introduces the reader to "Plant Science." What is Plant Science?

Plant Science encompasses several related fields within Botany such as physiology, pathology, ecology, evolution, breeding, genetics, bioinformatics, biotechnology, and genomics.

In this section, we trace some key history in the plant sciences. The timeline in Chapter 4 starts from the earliest forms of selection breeding for food, includes the classical work of Mendel in plant inheritance, and concludes with modern-day plant science. Key concepts in plant genetics, DNA, and mutation are covered in Chapter 5, which includes a "name your own mutant" exercise. The last chapter on edible mutants explores some plant mutants that you can find in your own grocery store.

Chapters in this section include:

Chapter 4 Mendel, McClintock, and the History of Plant Science

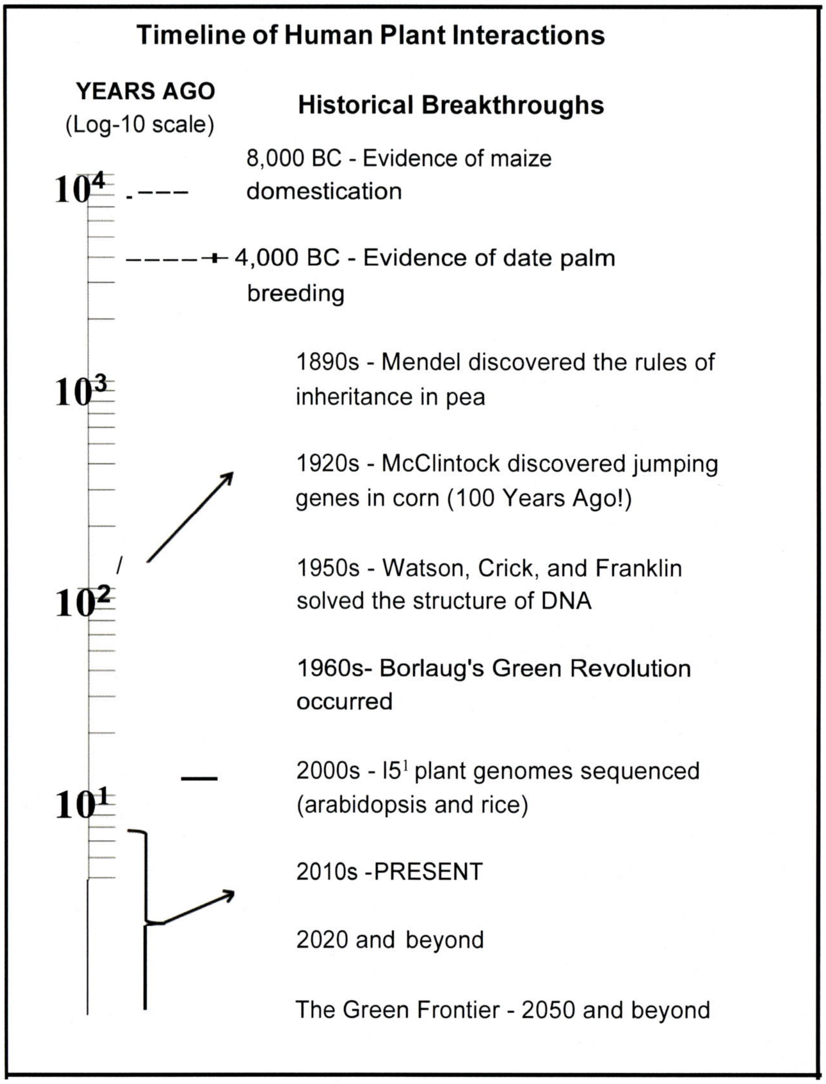

WHAT OTHER MAJOR EVENTS IN PLANT SCIENCE CAN YOU ADD?

HOW WOULD YOU FIND AN ANSWER TO THIS QUESTION?

WHERE ON THE TIME SCALE WOULD YOU PLACE YOUR ANSWER POINT?

From agriculture to genomics, through DNA

THE GATHERING OF THE SPATHES OF THE MALE PALM TREE.[1]

Section II: Figure 1. People breeding date palms, combining male (spathe) and female flowers, 2000-4000 BC. Image courtesy of gutenberg.org ebook History of Egypt at http://www.gutenberg.org/ebooks/17323.

Section II: Figure 2. The scientists who worked together to discover the double helix structure of the DNA molecule,1950s, Watson & Crick, and Franklin (These images may be freely reproduced and/or modified)..
Watson: wikimedia.org/windex.php?
CRICK:Wikipedia.org/wiki/Crick_Lecture#/m Image may be freely reproduced and modified. edia/File:FrancisCrick;Commons.wikimedia.org/wiki/File:Francis_Crick.png Photo by Marc Lieberman
Franklin:commons.wikimedia.org/wiki/File:Rosalind_CC-BY-SA.png.

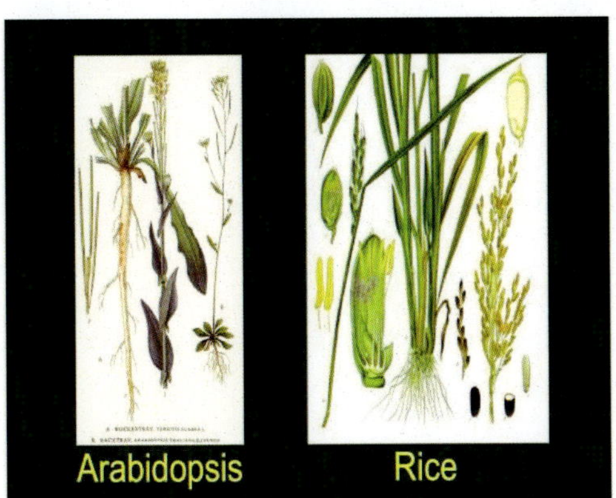

Section II: Figure 3. The first two plant species chosen for whole genome DNA sequencing, 2000s. Image courtesy of Wikimedia Commons at: https://commons.wikimedia.org/windex.php?curid=5969392 & https://commons.wikimedia.org/windex.php?curid=255589

Even before Mendel's famous experiments with the wrinkled peas, people were doing plant science in the form of selection or breeding, a practice that dates back 10,000 years (Section II Fig. 1). How did they do this? Three simple steps. Step 1: grow the plants, Step 2: observe the plants, and Step 3: get and replant seed from your favorite varieties. By selecting for desirable characteristics over and over, congratulations! you are breeding for the plant modifications you want. For a long time, plant scientists have been making new plant varieties, constantly selecting their favorites from the offspring (children) that grew from the seeds.

This activity is referred to as **selection breeding** and still very widely used all over the world to improve plant varieties.

Section II: Figure 4. Gregor Mendel. The monk who discovered the rules of heredity by studying how traits of the pea plant were passed on from parents (P, Parent generation) to children (called the F1 generation) and grandchildren (called the F2 generation).

A more advanced form of breeding involves what is called a **cross** - a deliberate mating of two different plants (the Mom and the Dad). A pioneer of plant breeding with crosses was the monk **Gregor Mendel**. He was fascinated by how characteristics passed on from parents to offspring in **family trees**. When family trees are drawn out, the drawings are called **pedigrees** and they record the **ancestry** and family structure.

The **characteristics** (or **traits**, also called **phenotypes**) that Mendel observed and recorded included:

- seed shape (round vs. wrinkled)
- flower color (purple (BB or Bb) vs. white (bb) as shown in Section II
- plant height (tall vs. dwarf)

His work resulted in several **Laws of Inheritance** (or **Mendelian genetics**) that apply to both plants and animals. Mendel's work became the foundation of the modern science of genetics.

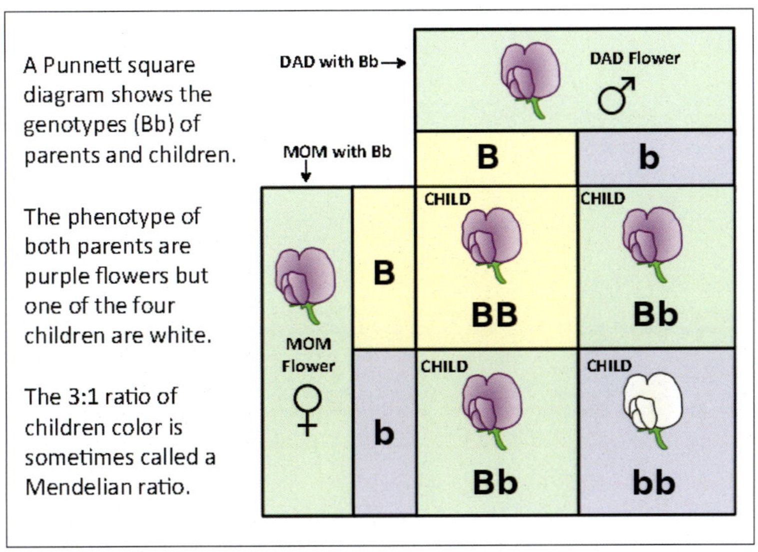

Section II: Figure 5. A Punnett square for one of Mendel's pea plant experiments. The parents are purple because they each have one big B gene. One of the four progeny is white, because they got the little b gene and white (bb) exhibits a recessive mode of inheritance. Modified from Madprime (Own work, CC By-SA 3.0, https://commons.wikipedia.org/w/index.php?).

Building on the foundation of genetic rules of inheritance, a famous scientist named **Barbara McClintock** went on to show how to use maize to understand even more basic principles and rules of plant genetics.

Section II: Figure 6. Image of Barbara McClintock as a young girl. Common.wikimedia.org/wiki/File:BarbaraMcClintock.jpeg Crop of File:McClintock family.jpg [[File:Barbara McClintock.jpeg|thumb|Barbara McClintock]]. Image may be freely reproduced and modified.

She is especially famous for her work on the discovery of **jumping genes**!

WHAT ARE JUMPING GENES?
DO WE HAVE JUMPING GENES (MOBILE DNA ELEMENTS) IN OUR DNA TOO?

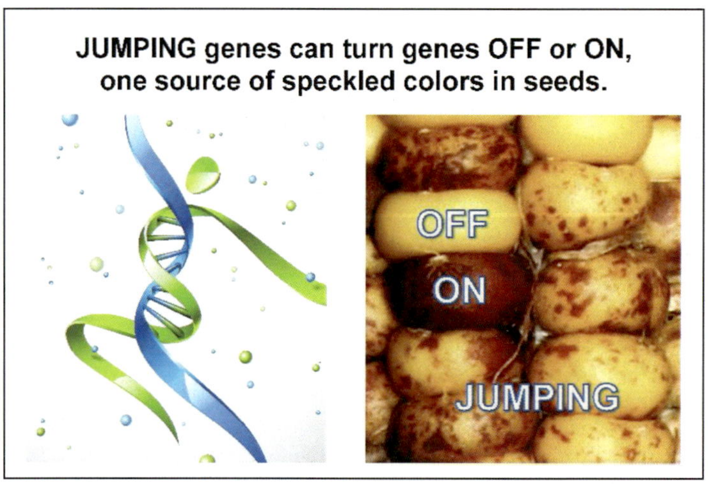

Section II: Figure 7. Jumping genes can switch genes on an off. Jumping genes are mobile DNA elements. If they hop in a gene it can get turned off (yellow). If they hop out, the gene can get turned back on (red). Image courtesy of HW Bass.

She also was a great cytogeneticist. **Cytogenetics** involves looking at the genetic material using a microscope. She counted and numbered the chromosomes of maize.

DO YOU KNOW HOW MANY CHROMOSOMES ARE IN A MAIZE CELL?

McClintock numbered the chromosomes from 1-10, but if you look in a cell, you may find 20! *What the Mendel?!* – Why 20 instead of 10? (Hint: see Chapter 9.)

Chapter 5 Plant Genetics: Working with DNA and Mutations

Many organisms reproduce by having children, also called **offspring** or **progeny**. Children or offspring from the same parents are called **siblings**. Brothers and sisters are siblings. Even the seeds on an ear of maize are siblings belonging to a family.

The reason that children resemble their parents is because they have inherited some genetic material from each parent. This genetic material is made of the chemical called ***DNA***. Do you know what the letters DNA stand for? The answer is de-oxyribo-nucleic-acid, a **double helix** molecule made of nitrogen-containing **bases** and **sugar-phosphate** backbone chains.

This genetic material controls how organisms grow and function, and also how they appear. The genetic term for the appearance is **phenotype**. For example, seed shape is a phenotype; height is a phenotype; seed color is a phenotype.

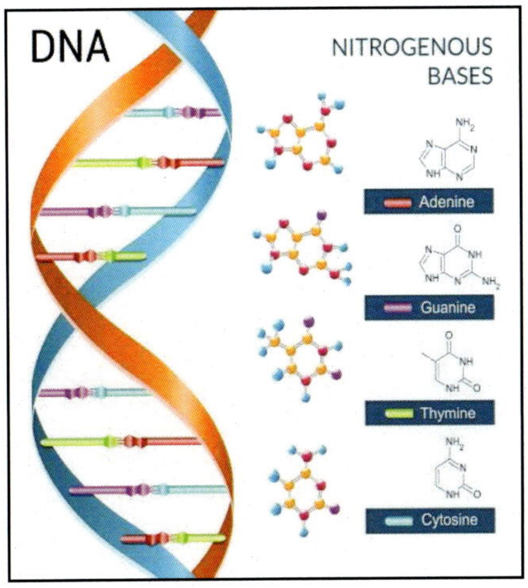

Section II: Figure 8. Diagram of DNA. The double-stranded chain of DNA bases is shown at the left. The individual bases and their names are shown at the right. Image from Dreamstime.com.

Phenotypes can "*run through a family*" because they are based on DNA inheritance.
Because DNA is passed along from parent to child, the children always resemble their parents. Squirrels have baby squirrels, oak trees have baby oak trees, and eagles have baby eagles.

Stable changes in small amounts of the DNA are called **mutations**. These too can be inherited and run through families.

DO YOU KNOW WHAT A MUTATION IS?

A **mutation** is most easily described as *"a heritable change in the* ***DNA*** *or the* genetic material *of an organism, tissue, cell, or even virus."*

The organism that has the mutation is called a **mutant**.

Mutations can change the appearance or phenotype of an organism. Examples of *phenotypes* and how they are used in plant genetics:

- **the *red color* in red seed maize is a phenotype**
- **the *purple color* in pea flowers is a phenotype**
- **the *shortness* of dwarf wheat is a phenotype**
- **the *striped fur coloring* on a zebra is a phenotype**
- **the *spotting* on a Dalmatian dog is a phenotype**

CAN YOU NAME SOME OTHER PLANT OR ANIMAL PHENOTYPES?

Plant biologists use MUTANTS to study biology and learn how the parts of organisms work.

How does that work? Here is an example to help you learn how mutants are very useful. You can think of a plant as being like a car; they both need all their parts working together to function correctly. In genetics, mutants are like a car but with one missing part. If a car is missing the brakes, it will not work correctly. If a car without brakes were a living organism, we would say that the brake-less car is a mutant. It would not be able to stop and we would learn from studying the brake-less mutant that brakes must help a car to stop moving. This is how we use mutants to learn how the parts of organisms work together. Similarly, what if a *mutant car* did not have a working horn, and was always silent. The quietness would be like a *phenotype*, a characteristic of the car. The horn-less mutant car would teach us that the horn is used to signal to other cars - HONK!!!

NOW IT'S YOUR TURN – PREDICT A CAR PHENOTYPE AND NAME A CAR MUTANT!

Suppose you have a "car mutant" that does not have a working gas pedal. What would be the phenotype of a car without a gas pedal? What might you name that car mutant?

Suppose you have a "car mutant" that does not have tires. What would be the phenotype of a car without tires? What might you name that car mutant?

Your turn – name a mutant. For a car mutant that does not have___________ (name a car part).

What would be the phenotype of a car without__________________? (your named part).

What might you name that car mutant?

DNA changes all the time and that is why there are so many different phenotypes seen in the world. This genetic variation is sometimes called **biodiversity**.

Not all mutations or changes in the DNA are bad or harmful. For example, a mutation in the gene that controls seed starch leads to sweet and wrinkled peas. The mutant produces sweet peas that dry down like raisins to appear wrinkled. This mutant is delicious while also being informative for the study of the rules of Mendelian genetics.

Did you know that the rubber tree plant produces the latex for making tires? This plant product is important for transportation because the latex is used for tires of cars, trucks, planes, and tractors. A bad mutation, for example, might reduce latex production. A good mutation, on the other hand, might make the rubber tree produce more latex or become more resistant to fungal diseases, such as the South American leaf blight.

Plant mutants are useful for many reasons:

- They give clues about how genes help plants grow.
- They provide natural genetic variation in groceries.
- They provide natural genetic variation for flowers and ornamentals.
- They provide natural genetic variation in industrial plant products.
- They provide natural genetic variation in medicinal plants.

CAN YOU NAME ANY PLANTS THAT COULD BE IN TWO OR MORE OF THESE FIVE CATEGORIES?

Chapter 6 Edible Mutants: Dwarf Wheat, Seedless Fruits, and Sweet Corn

EDIBLE DWARFS

Did you know that much of your bread, pasta, and rice comes from dwarf or semi-dwarf mutant varieties? Dwarfing in this case is a plant phenotype associated with thick stems supporting stronger and more numerous stalks. For grain plants, such as wheat or rice, these multi-stalked varieties make more cereal grains.

SEEDLESS FRUITS

Did you know that many seedless fruits are developed on purpose by plant breeders and geneticists? Seedless fruits do not contain mature seeds, making them easier and more convenient to eat. To learn more about all the different ways to make seedless fruit, visit: https://en.wikipedia.org/wiki/Seedless_fruit

Common varieties of seedless fruits we love to eat include some varieties of watermelons, tomatoes, grapes, bananas, oranges, lemons, and limes.

Seedless watermelon fruits are produced on **triploid** plants, whose three sets of **chromosomes** (see chapter 10) cause the seeds to be genetically unstable and sterile – so the seeds do not form their full normal size and appear tiny, if at all.

Section II: Figure 9. Seedless watermelon. Seedless watermelon comes from genetic crosses that produce triploid plants. These triploids can grow and make fruit, but not seeds. Image by Scott Ehardt - Own work, Public Domain. Image courtesy of Wikimedia commons at https://commons.wikimedia.org/w/index.php?curid=65565.

OTHER SEEDLESS FRUITS

In addition to watermelon, seedless varieties of other fruits such as some types of grapes, oranges, and most types of banana have been bred for us to eat.

Section II: Figure 10. Bananas. Photographed by Adrian Pingstone in March 2004 and released to the public domain. {PD}. Located at Wikimedia.com https://commons.wikimedia.org/wiki/File:Banana.arp.750pix.jpg

Section II: Figure 11. Seedless Orange. The navel orange is reportedly originally from a single Brazilian tree found in the 1800s. Image from Wikimedia Commons. at: Orange_seedless **Challiyan at malayalam Wikipedia. https://upload.wikimedia.org/wikipedia/commmons/4/47/orange_seedless.jpg**

SWEET CORN IS A DELICIOUS MUTANT!

DID YOU KNOW THAT SWEET MAIZE IS A MUTANT WITH A WRINKLED SEED PHENOTYPE?

These seeds are wrinkled and shrunken because one or more of their starch-producing genes are broken (mutant). As a result, they cannot fill the seed with starch. Instead they fill the seed with small soluble sugars.

This ear of maize has both normal (full seed) and mutant (wrinkled sweet maize) seed. The genetic name of this mutant is *sugary*.

Fun Fact: Sweet maize mutants are similar to Mendel's famous wrinkled pea mutants, also known as sweet peas!

Section II: Figure 12. An ear with kernels that are either normal (full sized) or mutant (winkled). These raisin-like seeds are actually caused by a mutation in the gene called "*sugary1*". Image source courtesy of MaizeGDB at https://images.maizegdb.org/db_images/Variation/mgn/7101_3161_2557_17.jpg.

DID YOU KNOW MAIZE HAS MANY USES THAT ARE NOT RELATED TO FOOD?

Those are described in Section III on non-edible uses of maize.

III. NON-EDIBLE USES OF CORN

Introduction

This section introduces the reader to how maize can be used for many purposes beyond those of human food or animal feed. Examples range from bioplastics made from seed extracts to folk medicines made from maize silk. In this section, we discuss how maize, like other plants, is able to capture solar energy and store it in little molecular batteries made of sugar and chains of sugar like starch. In some cases, those sugars can be converted to *ethanol fuel* to power automobiles, planes, and boats. In other cases, maize starch can be used to make valuable products such as earth-friendly bioplastics additives for better textiles.

Chapters in this section include:

Chapter 7 Biofuel, Ethanol Fuel Can be Made from Corn

CAN YOU EXPLAIN WHY PEOPLE AND OTHER ORGANISMS NEED TO EAT FOOD?

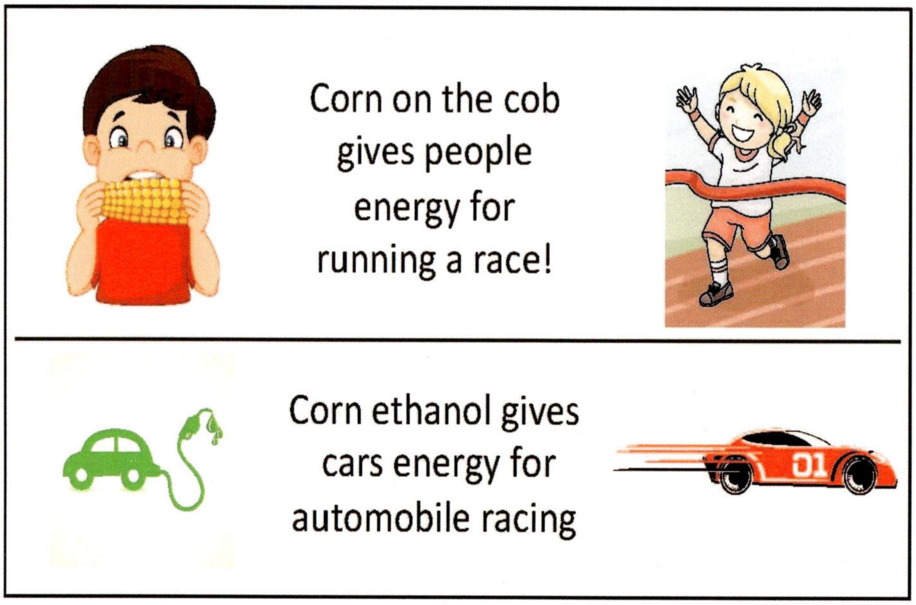

Section III: Figure 1. Corn is a source of energy for people and cars.

People eat plants for **energy** to live and grow. These plants provide a biological type of **fuel** which can be used to power us, but also machines. The plant sugars (**carbohydrates**) can be converted into what is called biofuel.

HAVE YOU EVER SEEN THE WORD "ETHANOL" AT THE GAS PUMPS?

If you have, it is because ethanol can be made from maize or other plant sugars and added to gasoline to power cars.

Section III: Figure 2. Some types of gasoline include the plant-based fuel ethanol. A common source of plant ethanol is corn. Image from Shutterstock.

WHERE DO PLANTS GET THEIR FUEL? THEY MAKE IT THEMSELVES!

The plant sugars, such as **glucose**, store energy that originally came from the sun using a process called **photosynthesis**.

Photosynthesis is the process by which green plants and some other organisms use **sunlight, carbon dioxide,** and water to synthesize sugar and release **oxygen**.

In this way, we can think of living plants as biological solar panels to make biofuels for gasoline and other types of fuels.

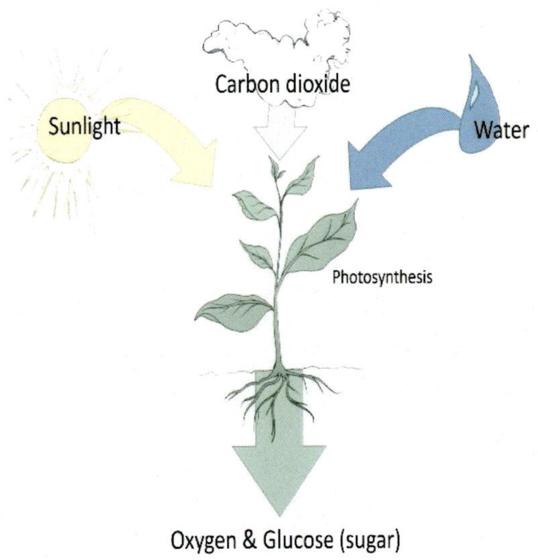

Section III: Figure 3. Diagram of photosynthesis, the process of combining energy from the sun with carbon dioxide and water produce sugar and oxygen.
Diagram courtesy of Allison Jevitt.

WHY DOES SUGAR GIVE US ENERGY?

Sugars store energy in **chemical bonds**. When these bonds break – through digestion in our stomach or burning in the car engine – they release energy!

DID YOU KNOW THERE ARE LOTS OF PLANT-BASED FUELS?

CAN YOU NAME SOME PLANTS USED TO MAKE ETHANOT FUEL OR BIODIESEL?

Some plants used as fuels include grasses (sugarcane, switchgrass, maize), seed oil crops (soybean, carinata, canola, palm oil), poplar trees, and algae.

DID YOU KNOW CORN GRAIN USED FOR ETHANOL FUEL CAN POWER A CAR?

CANOLA AND CARINATA CAN POWER A PLANE.

Section III: Figure 4. Fuel from the plant carinata. Image courtesy of Biodiesel.org.

POPLAR TREES CAN POWER INDUSTRIAL PRODUCTS

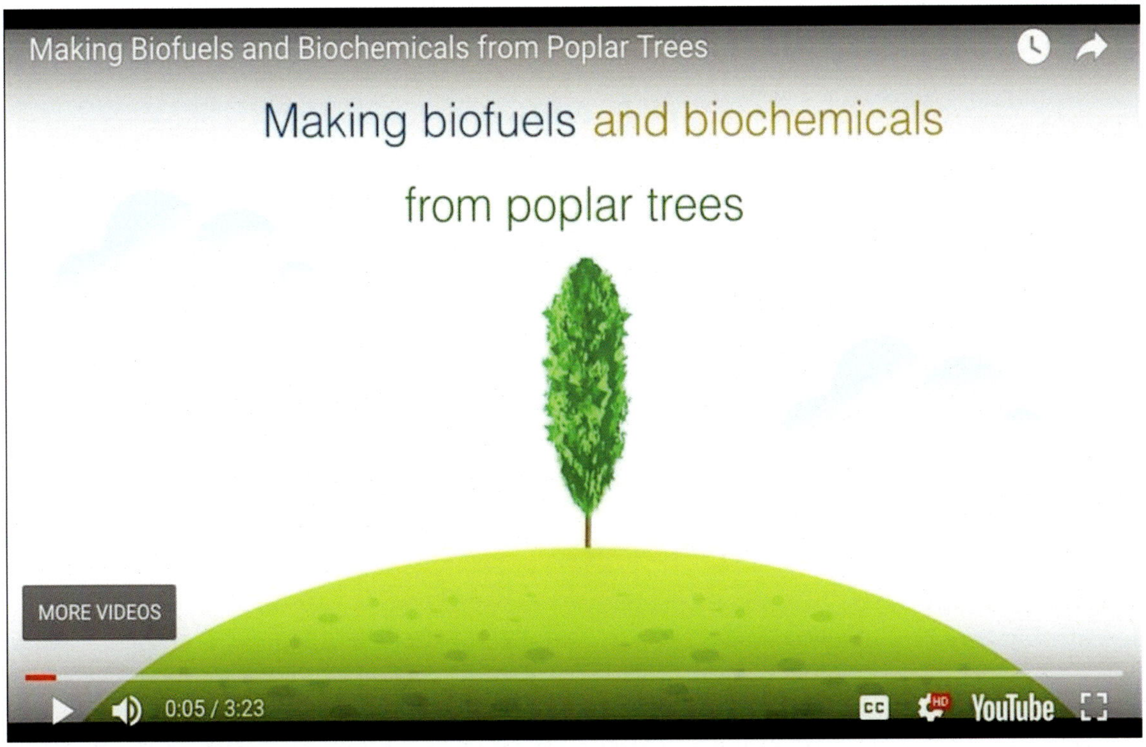

Section III: Figure 5. Educational video about plant biofuels and biochemicals. Produced by a team of plant scientists at Washington State University Extension, University of Washington, and Advanced Hardwood Biofuels Northwest. From YouTube video xQXRFGu4JXA.

CAN YOU FIND THE FOLLOWING ITEMS RELATED TO STEM IN THIS VIDEO?

Science ____________________

Technology ____________________

Engineering ____________________

Math ____________________

Research Teams ____________________

Biodegradable Plastics ____________________

Chapter 8 Bioplastics and Industrial Starch Can be Made from Corn

Plastics that are made from natural plant products are called **bioplastics**.

One of the most common bioplastic materials comes from a chemical called **polylactic acid**, abbreviated **PLA**. **PLA** can be made from ingredients isolated from kernels of maize.

Bioplastic, unlike regular plastic, is **biodegradable**.

Biodegradation refers to the breakdown, rotting, or eating of materials by microbes such as bacteria, fungi, or other biological means.

Bioplastics are viewed, therefore, as environmentally friendly.

Bioplastics made from maize are used in all kinds of items, including:

- plastic utensils
- packing peanuts
- bags and wrapping materials

DO YOU HAVE ANY BIOPLASTICS IN YOUR HOUSE?

Section on III: Figure 6. Biodegradable plastic utensils. Image courtesy of SokheChapke Publishing. Image may be freely reproduced and modified.

Section III. Figure 7. The clear wrapping around these flowers is made in part from PLA, a bioplastic and the wrapping is made from natural ingredients from maize. Image courtesy of SokheChapke Publishing. Image may be freely reproduced and modified.

DID YOU KNOW THAT MAIZE STARCH IS NOT ONLY A FOOD, BUT ALSO HAS MANY INDUSTRIAL USES?

Corn starch has many names.

It is also called *maizestarch*, *maizeflour* or *maize starch*. Corn starch is derived from the kernel of the maize. Some factories are designed to produce maize starch for textiles and other industries. Starch for this purpose is called "industrial maize starch" to distinguish it from food starch.

Corn starch is very valuable in the laundry and dry-cleaning industries.

WHY DO PEOPLE USE STARCH TO CLEAN CLOTHES?

For hundreds of years, people have been using plant starch to clean fabrics because starch binds to dirt and grime. The dirt sticks to the starch instead of the clothes and can be rinsed away after washing for bright clean clothes.

Corn starch can also improve the cloth-dyeing process used to make colorful fabrics and towels.

There are so many uses of maize! No wonder people love it! Another thing people love about maize is its genetic diversity and available collection of famous mutants used for science. The next section features some of the most beloved mutants of maize.

IV. MAIZE MUTANTS

Introduction

This section introduces the reader to the wonderful world of maize mutants and genetics. As introduced in Chapter 5, mutations in DNA result in individual mutants. These mutations are heritable, passed down from the parent to the child. These chapters will introduce the reader to basic genetic concepts that describe how the genetic material is organized into long DNA molecules that make up the ***chromosomes***. Maize has 10 different chromosomes, and the location of genes and mutations have been mapped to individual chromosomes using genetic crosses. The technique of making a ***genetic cross*** in maize is described in Chapter 11, using illustrated picture guides and photos of students carrying out ***controlled pollinations***. The last chapter in this section highlights some of the students' favorite mutants, each with its own biological story and mystery to solve.

Chapters in this section include:

Ch9 The DNA, Chromosomes, and Genome of Maize
Ch10 How to Make a Genetic Cross of Maize
Ch11 The Crazy and Beautiful Mutants of Maize

Chapter 9 The DNA, Chromosomes, and Genome of Maize

Genes are the basic units of inheritance that are passed along from a parent to offspring (children).

Genes determine most of an organism's characteristics, such as height, color, and other phenotypes.

DO YOU KNOW THAT A MAIZE PLANT HAS OVER 40,000 GENES?

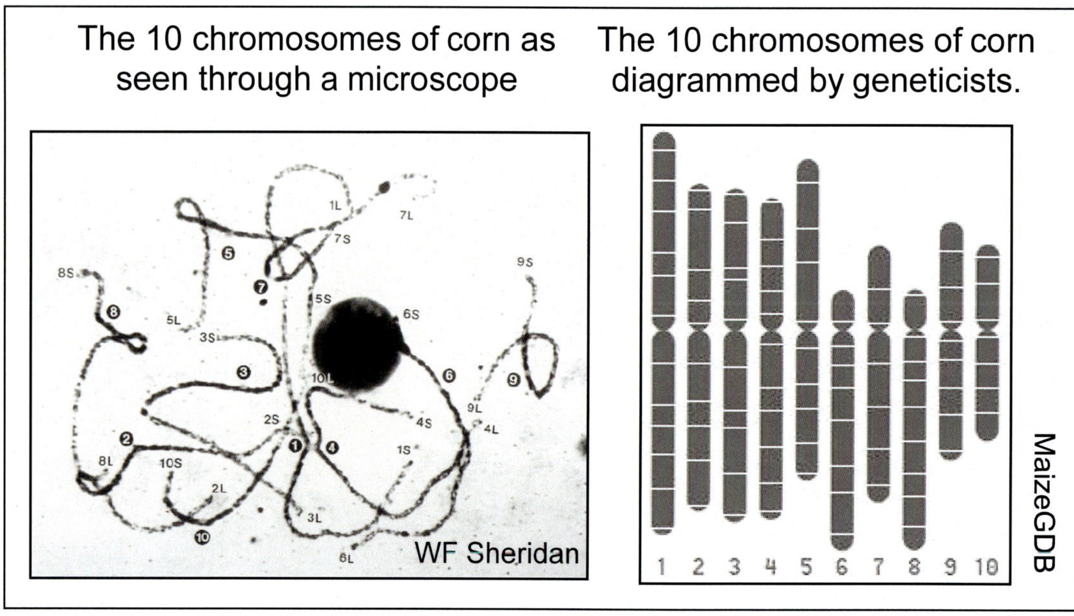

Section IV: Figure 1. Chromosome spread and karyotype of maize. The 10 chromosomes of maize (left) have short (S) and long (L) arms and a big black sphere called a nucleolus, attached to chromosome 6. Image courtesy of The diagram (right) shows the 10 chromosomes and "genetic bins" – similar to city blocks. Images are from Dr. William F. Sheridan (left). The maize pachytene spread was prepared by John Stout and the negative used to prepare the figure was kindly provided by Ronald L. Phillips. The image on the right is courtesy of MaizeGDB www.maizegdb.org/bin-viewer.

For most organisms, there exist two matching sets of chromosomes in each cell, one set from each parent. For example, although maize has 10 different chromosomes, each cell has 20 chromosomes, 10 from each parent. This double genetic material is referred to as a "**diploid**" genetic state. Similarly, humans have 23 chromosomes, but our cells have 46 because there are 23 from the mother plus 23 from the father.

Table of chromosome numbers for several different organisms

Organism	Haploid chromosome No.	Diploid Chromosome No.
Human	**23**	**46**
Dog	**39**	**78**
Maize	**10**	**20**
Bread Wheat	**21**	**42**
Pea	**7**	**14**
Orange	**9**	**18**

WHICH HAS MORE CHROMOSOMES, MAIZE OR ORANGE?

WHICH HAS MORE CHROMOSOMES, SWEET CORN OR BLUE CORN?

Answer: The second question is a trick question! Both maize types have a haploid chromosome number of 10. But the genes on those chromosomes may be slightly different. For example, one of the genes that can result in sweet maize is a mutant called sugary1 found on maize chromosome 4 (Section IV, Fig. 1). Plant geneticists have been mapping genes and mutations to chromosomes for over 100 years. Learn how to make maize genetic crosses in the next chapter.

Chapter 10 How to Make a Genetic Cross of Maize

In plants, a **genetic cross** is used to produce new seeds from known parents. Like animals, plant crosses involve two parents, a male and a female.

When the **pollen** from the male and **eggs** from the female are combined, **fertilization** occurs, forming an offspring. In fertilization, each parent cell adds one set of chromosomes to the **zygote**, which develops into a seed. The pollen and egg cells of plants are haploid (see Chapter 9) and the offspring (children, seed) are diploid. A maize cross produces an ear of corn in which the seeds (kernels) can be thought of as children from one family.

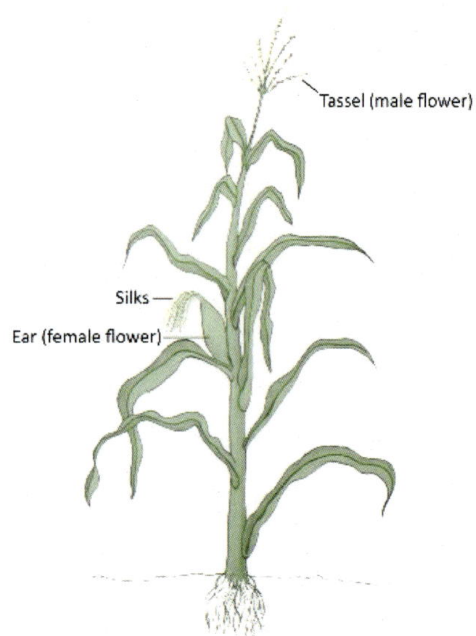

Section IV: Figure 2. The male and female flowers of maize occur on the same plant but in different locations. Diagram courtesy of Allison Jevitt.

In maize, the male flower, which are like the dads, are found in the **tassel**, at the top of the plant.

In maize, the female flowers, which are like the moms, are found in the **ear**, along the stalk of the plant.

The location of the tassel and ear are shown (Section IV, Fig. 2).

Three Simple Steps for Making a Maize Cross

Step 1. Bag the tassels to collect pollen (male)

Step 2. Pour pollen on silks of the ear (female)

The cross is complete!

Step 3. Label your crossing bag and staple it over the ear for later harvest.

Section IV: Figure 3. How to make a maize cross. Demonstrated by FACE students in Florida, Photos courtesy of Jonathan Doster (dosterphoto.com).

HOW TO MAKE A MAIZE CROSS – YouTube Video

Section IV. Figure 4. Video showing how to make a corn cross. Demonstrated by Dr. Bass and Julian Dennis with help from the Young Scholars Program (YSP) students. Link to video at https://goo.gl/KeH9eJ.

Chapter 11 The Crazy and Beautiful Mutants of Maize

As described in Chapter 5, plant biologists use mutants to help them study how genes work and affect the appearance (phenotypes) of organisms. A large and extraordinary collection of maize mutants have been mapped to chromosomes by maize genetics researchers for nearly 100 years. The photographs in this chapter are from the *Maize-10-Maze* project produced by FAMU and FSU researchers with support from Plant Genome Research Program of the National Science Foundation. Photography by Jonathan Doster (http:// www.jdosterphoto.com).

This mutant is called ***lazy plant1 (la1)***

WHAT IS THE PHENOTYPE OF THIS MUTANT?

Section IV: Figure 5. The *lazy1* (*la1*) maize mutant. These mutant plants grow along the ground. For more information: https://www.maizegdb.org/gene_center/gene/la1.

This mutant is called *Corngrass1 (Cg1)*

WHAT IS THE PHENOTYPE OF THIS MUTANT?

Section IV: Figure 6. The *Corngrass1* (*Cg1*) maize mutant. Changes the single stalk into a bushy plant with lots of side branches. For more information: https://www.maizegdb.org/gene_center/gene/cg1.

This mutant is called *Knotted1 (Kn1)*

WHAT IS THE PHENOTYPE OF THIS MUTANT?

Section IV: Figure 7. The *Knotted1* (*Kn1*) maize mutant. This knotted1 mutant of maize show out of control growth. For more information: https://www.maizegdb.org/gene_center/gene/kn1.

This mutant is called *Rolled leaf1 (Rld1)*

WHAT IS THE PHENOTYPE OF THIS MUTANT?

Section IV: Figure 8. The *Rolled leaf1* (*Rld1*) maize mutants. The mutant leaves fail to unroll properly during development. Picture courtesy of Karen Graffius, FSU Mission Road Research Facility.

For more information: https://www.maizegdb.org/gene_center/gene/rld1.

This mutant is called *Colored1-marbled (R1-mb)*

WHAT IS THE PHENOTYPE OF THIS MUTANT? WHY WAS IT NAMED MARBLED?

The color pattern is caused by a transposon (jumping gene!) as described in Chapter 4. When the transposon (also called a mobile DNA element) jumps out of the *R1* gene, it can reactivate the R1 gene, which makes colored pigments and the dark marble-like patches as seen below.

Section IV: Figure 9. The *R1-marbled* (*R1-mb*) maize mutant.
Pictures of seeds with the R1-marbled mutant seeds. For more information about the R1 gene, see https://www.maizegdb.org/gene_center/gene/r1.

This mutant is called *white luteus5 (wlu5)*

WHAT IS THE PHENOTYPE OF THIS MUTANT?

This leaf color mutant is **albino**, which means that it lacks pigmentation. The mutant seedling (white and its normal siblings (green are shown.

These mutants only grow as seedlings because they cannot make their own food through **photosynthesis** as described in Chapter 7 (Section III, Figure 3. As seedlings, they will live for a little while by using the food reserves from the seed.

Section IV: Figure 10. The *white luteus5* (*wlu5*) maize mutant. Picture of seedlings showing the recessive phenotype of this albino plant mutant. For more information about the wlu5 gene, see https://www.maizegdb.org/gene_center/gene/wlu5.

This mutant is called *Zebra crossbands8 (Zb8)*

WHAT IS THE PHENOTYPE OF THIS MUTANT?

This mutant exhibits a red striping or banding pattern across its leaves.

It looks like zebra stripes!

Section IV: Figure 11. The *Zebra crossbands8* (*Zb8*) maize mutant. Picture of maize leaves showing the mutant striping. For more information about the Zb8 gene, see https://www.maizegdb.org/gene_center/gene/zb8.

This mutant is called *dwarf3 (d3)*

WHAT IS THE PHENOTYPE OF THIS MUTANT?

HAVE YOU EVER HEARD OF DWARF PLANTS BEFORE?

(Hint: The Green Revolution, Chapter 2)

Section IV: Figure 12. The *dwarf3* (*d3*) maize mutant. Picture of maize dwarf3 mutant plant in the field. For more information about the d3 gene, see https://www.maizegdb.org/gene_center/gene/d3

V. ARTS AND CRAFTS WITH CORN

Introduction

This section introduces the reader to the art, poetry, and culture of corn. Chapter 12 includes maize-inspired poetry by the authors, including poems that celebrate the pioneering work of Nobel Laureate, Barbara McClintock, using maize to discover jumping genes. Chapter 13 describes the importance of maize to the Mayans, Aztecs, and other Native Americans. They worshiped maize and had maize deities (maize gods) going as far back as 200 AD. One tradition, called the Green Corn Ceremony, is still celebrated in this century by Seminoles and other Native American tribes. Finally, Chapter 14 encourages students and teachers to explore the crazy world of maize mutants by ordering seeds and growing their own.

Chapters in this section include:

Chapter 12 Poetry and Art of Corn

BABARA McCLINTOCK: FEARLESS PIONEER OF CYTOGENETICS AND MAIZE JUMPING GENES!!!

By Kome Onokpise and Hank Bass

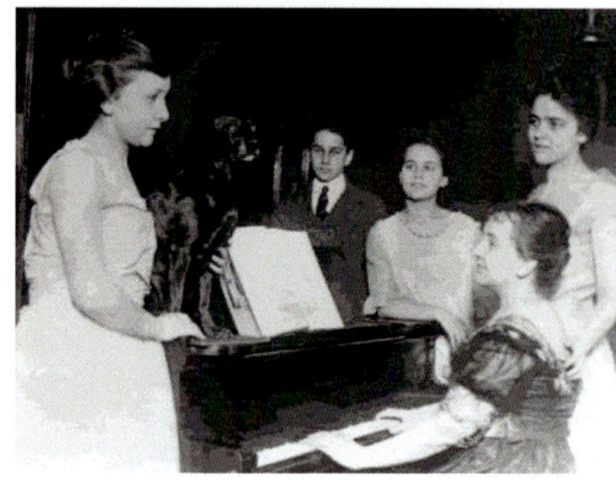

Section V: Figure 1. Barbara McClintock with family. From left to right: Mignon, Malcolm Rider "Tom," Barbara, Marjorie, and Sara (at the piano). Image courtesy of Wikimedia commons at https://commons.wikimedia.org/w/index.php?curid=22971541.

The year was 1902 when you were born
You had two sisters and one brother
you developed independent thoughts
Led to your tenacity for landmark research in Maize Genetics

Fearless mother of **cytogenetics**
Fusing cytology and genetics was your great idea
Your work on the gene movement in maize chromosomes
Enabled us to understand maize kernel variation (coloration)

The **transposable elements** "TE" or Jumping
Genes are the basis for our understanding of the
Controlling elements that regulate genes in living organisms
The knowledge now used in genomic research

With corn or maize as the model organism
Discovered the **Activator** (*Ac*) and **Dissociator** (*Ds*)
Transposon system, a tool now used to produce mutant plants
For the characterization of gene function

So now we know more about maize genetics and mutants
Why one is an albino
Another has Knotted stems and leaves
And yet another grows CRAZY as if it were LAZY

And you became a Nobel Laureate at the age of 81
The first woman to earn an unshared Nobel Prize
A great role model for young and old scientists,
Leaders of the next green revolution to feed 10 billion people

Through your scientific research and love of maize
We can share the science of plant genetics with young students to learn about *Jumping Genes Crazy Lazy Corn* salutes and thanks Dr. Barbara McClintock!

Section V: Figure 2. Barbara McClintock doing cytogenetics on maize. Looking at chromosomes in 1947 using a microscope, Dr. McClintock was a pioneer plantgenome researcher. Image from Smithsonian Institution/Science Service, restored by Adam Cuerden. Image courtesy of Wikimedia commons at https://commons.wikimedia.org/w/index.php?curid=22971541.

What is **FACE?**

By Kome Onokpise and Hank Bass

Forestry and Conservation Education FACE was
Developed for middle and high school students
To learn about the science of natural resources and their conservation

The trees, the wildlife, the insects, and all the
Other plants How do they look?
Any fire in the forest? How do we manage that?

So why did FACE work with Maize?
Isn't maize the same as corn? Yes, it is also a
Plant! So the students learn about the
Science of plants, including trees and maize

And what about the Maize?
Yes, the FACE students go the Maize Genetics Lab
And then to the maize fields to prepare the *Maize-10-Maze*
A walk through maize genome with mutants from all 10 chromosomes

They find these different types of maize
Looking so different like trees in a forest
With lazy, striped, dwarf, and knotted maize mutants
FACE to face with the Maze-O-Maize

Section V: Figure 3. FACE students. Students working in the summer maize DNA laboratory (top two pictures) and in a group photo (bottom picture) with Dr. Bass at Florida State University.

Chapter 13 Cultural History of Maize

Like other Mesoamerican people, the traditional Mayans recognized in their staple crop, maize, a vital force with which they strongly identify. This is clearly shown by their mythological traditions.

According to the 16th-century Popol Vuh, the Hero Twins have maize plants for alter egos in a world where people come from maize. The discovery and opening of the Maize Mountain - the place where the maize seeds are hidden - is still one of the most popular of Mayan tales. In the Classic period (200-900 AD), the maize deity became a cultural hero. "People of Corn, A Mayan Story" retold by Mary-Joan Gerson, is from

http://wakinguponturtleisland.blogspot.com/2013/09/maya-influence.html

Section V: Figure 4. Cultural and historical example involving maize.
Three Sisters as featured on the reverse of the 2009 Native American U.S. dollar coin. Source: https://en.wikipedia.org/wiki/Three_Sisters_(agriculture)

DID YOU KNOW THAT THE MAIZE GOD IS NAMED CENTEOTL?

Section V: Figure 5. Centeotl, one of the maize gods.

https://commons.wikimedia.org/wiki/File:Cinteotl_Borgia.jpg

HAVE YOU EVER HEARD OF THE GREEN CORN FESTIVAL? DID YOU KNOW THAT IT IS STILL BEING CELEBRATED?

The historic Green Corn Festival is also called the Green Corn Dance or Green Corn Ceremony. The celebration and religious ceremony is a dance held by many different Native American tribes, including the Seminole tribes native to Florida and Oklahoma.

The festival typically lasts for three days, with the opening day determined by when the local maize is ripe. These festivals can be any time from May to October, determined by the "Keepers of the Faith." In this tradition, maize is not to be eaten (because it is still green) until the Great Spirit has been given proper thanks.

Section V: Figure 6. Corn celebrations. The Green Corn Dance in 1861. Right: The Green Corn Dance in 2014. Image sources: Left image from By George Catlin, from the Paul Mellon Collection, courtesy of Wikimedia commons curid=50295913.

At the end of every day, the people feast on slabs of beef, maize soup, beans, and squash. The tribes celebrate a good harvest and enjoy various foods made from maize, including tortillas, maize bread, and maize soup. More information about the Green Corn Festival is available from "People of Corn: A Mayan Story" (1995) by Mary-Joan Gerson and from http://teacherlink.ed.usu.edu/tlresources/units/Byrnescelebrations/corn.html.

Chapter 14 How to Find and Grow Your Own Mutants

DID YOU KNOW THAT YOU CAN GET AND GROW YOUR OWN MUTANT SEEDS OF MAIZE?

Your science teacher can request seeds from the **Maize Genetics CO-OP.**

What does the Maize Seed Co-op do?

The Maize Genetics Cooperation Stock Center is operated by the **United States Department of Agriculture/Agricultural Research Service (USDA/ARS),** located at the University of Illinois,
Urbana/Champaign and integrated with the National Plant Germplasm System (NPGS). The maize seed co-op serves the maize research community by collecting, maintaining and distributing seeds of maize genetic stocks.

The seed co-op and catalog are online at http://maizecoop.cropsci.uiuc.edu/

Maize Genetics Cooperation Stock Center

Section V: Figure 7. Examples of three mutants obtained from the Maize Co-Op Stock Center. The mutants shown here are Left – *dwarf3*; Middle – white *luteus5*; and Right – *Knotted1*.

Need some advice for using maize mutants in the classroom or for student projects?

For additional information or advice on how to grow or obtain mutant seeds, please contact:

Dr. Bass (bass@bio.fsu.edu) at Florida State University

Dr. Onokpise (oghenekome.onokpise@famu.edu) at Florida A&M University

Section V. Figure 8. Plant scientists help people. Shown here is Mr. Zadarreyal "DJ" J. Wiggins. He earned his Masters degree in Plant Science from FAMU working with Dr. Onokpise and Dr. Bass on maize genomics. In this photo, he was an Agricultural Extension Agent – Plant Science Specialist, Gadsden County, FL. Extension agents help people by educating farmers and other members of the public on matters related to agriculture. He is currently in graduate school working to earn a PhD.

Glossary

Activator/Dissociator (Ac/Ds). A type of jumping gene in maize, discovered by Barbara McClintock. These are mobile DNA elements also called transposons or transposable elements (TEs).

Agricultural Mechanization. Using machinery for agriculture and farming, replacing animals, and increasing farm worker productivity.

Ancestry. The record of parents in families and family trees.

Base (of DNA). Part of the DNA molecule, one of the building blocks of DNA. The four bases of DNA are abbreviated A, C, G, or T.

Bass, Henry W. A plant geneticist and university instructor and research scientist at Florida State University in Tallahassee, FL, USA.

Biodegradation. Refers to the breakdown, rotting, or eating of materials by microbes such as bacteria, fungi, or other biological means.

Biofuel. A fuel made directly from living organisms and their products instead of from prehistoric sources such as fossils.

Biological Diversity. Also called biodiversity, refers to the full variety of living organisms in the whole world or in a particular region of the world.

Borlaug, Norman. An American agronomist and humanitarian who saved people from starvation by increasing food production through plant breeding and farming practices.

Carbohydrate. A sugar, made up of carbon, hydrogen, and oxygen atoms and produced by plants during photosynthesis.

Centeotl. A maize god and son of the earth goddess according to the Aztec mythology. The name translates into "deity/god of dried maize still on the cob."

Chemical Bond. A type of connection between atoms, ions, or molecules resulting in chemical compounds. These bonds can be weak or strong. Strong bonds can release energy when broken.

Chromosome. The long DNA molecules in the cells that contain the genetic material. They are replicated and passed on during cell division and reproduction, providing the basis of inheritance.

Controlled Pollination (see Genetic Cross)

Corn Starch. A starch (a chain of glucose sugars) derived from corn.

Crop Domestication. The process by which humans conduct selection and breeding of plants and crops for beneficial uses.

Cytogenetics. The study of chromosomes and genetics by looking at cells with the aid of microscopes.

Diploid. Refers to the two matching sets of chromosomes in each cell, one set from each parent. For example, although maize has 10 different chromosomes, each cell has 20 chromosomes, 10 from each parent.

DNA. Deoxyribonucleic acid, a double helix molecule made of nitrogen-containing bases and sugar-phosphate backbone chains. Primary carrier of genetic information controlling how organisms grow and function, and also how they appear.

Ear. The seed or grain-bearing tip part of the stem of a cereal plant, such as wheat or maize.

Egg. The female reproductive cell or gamete that contains one set (haploid) of chromosomes.

Energy. Power derived from the utilization of physical, chemical and natural resources to provide light, heat and/or to work machines.

FACE. Forestry and Conservation Education. FACE was developed for middle and high school students to learn about the science of natural resources and their conservation

Family Tree. (see Pedigree)

Fertilization. When the pollen from the male and eggs from the female are combined to form a diploid zygote that develops into an offspring. In fertilization, each parent cell adds one set of chromosomes.

Food Security. A condition related to the availability of food supply, groups of people such as ethnicities, racial, cultural and religious groups, as well as individuals' access to it.

Gene. The basic unit of inheritance located in a specific locus of chromosomes it is transferred from parent to offspring and thought to determine the characteristics of the offspring.

Genetic Cross. Production of new seeds from known parents. Like animals, plant crosses involve two parents, a male and a female.

Germplasm. The total of all hereditary material in a single species. In plants, germplasm collections are seed banks.

Jumping Gene. (see Transposon)

Locus. Site or place on a chromosome where a gene is located.

Maize Genetics Co-op. Stock Center, operated by the United States Department of Agriculture/ Agricultural Research Service (USDA/ARS), located at the University of Illinois, Urbana/Champaign, and integrated with the National Plant Germplasm System (NPGS). The maize seed coop serves the maize research community by collecting, maintaining and distributing seeds of maize.

Maize-10-Maze. A public outreach activity designed to share the wonder of plant genetics with students (https://www.bio.fsu.edu/bass/maizenucleosome/outreach/).

Mayan. Refers to various indigenous peoples of Mesoamerica and northern Central America.

McClintock, Barbara. (June 16, 1902 – September 2, 1992) was an American scientist and cytogeneticist who was awarded the 1983 Nobel Prize in Physiology or Medicine.

Mendel, Gregor. Austrian Monk who founded modern day genetics and formulated the laws of inheritance

Mendelian Genetics. Laws of inheritance that apply to plants, animals, and fungi.

Mutant. An organism that carries a heritable variation as a result of mutation

Mutation. A heritable change in the DNA or the genetic material of an organism, tissue, cell, or even virus.

Nobel Peace Prize. Awarded for outstanding contributions in peace. According to Nobel's will, the peace prize "shall be awarded to the person who in the preceding year shall have done the most or the best work for fraternity between nations, for the abolition or reduction of standing armies and for the holding and promotion of peace congresses."

NSF (National Science Foundation). A United States government agency that supports fundamental research and education in all the non-medical fields of science and engineering. Its medical counterpart is the National Institutes of Health.

Onokpise, Oghenekhome U. A plant geneticist and university instructor and research scientist at Florida Agricultural and Mechanical University in Tallahassee, FL, USA.

Pedigree. A diagram showing the ancestral relationships and transmission of genetic traits over several generations in a family.

Phenotype. Observable characteristics or traits (physical or functional appearance) of an organisms resulting from the interaction of its genotype with the environment.

Photosynthesis. The process by which green plants and some other organisms use sunlight, carbon dioxide, and water to synthesize sugar and release oxygen.

Ploidy. Refers to the number of sets of chromosomes in a cell or in the cells of an organism; haploidy = 1 set; diploidy = 2 sets, triploidy = 3 sets, and so on.

Pollen. Fine powdery substance, generally yellow and microscopic, discharged from the male part of a flower or cone. The pollen carries the genetic material of the male parent.

Polylactic Acid (PLA). Chemical compound from which the most common bioplastic materials are produced.

Selection Breeding. An activity in which plant scientists select for desirable characteristics over and over from parent and offspring to produce new and improved varieties.

Semi-Dwarf. A plant phenotype associated with thick stems supporting stronger and more numerous stalks. For grain plants, such as wheat or rice, these multi-stalked varieties make more cereal grains.

Tassel. The male flowers, which are found at the top of corn (maize) plant.

Trait (see also Phenotype). The heritable characteristic of an organism.

Transposon (Transposable Element, TE). DNA sequence capable of moving (transposing) from one location to another in a genome, also called a jumping gene.

Triploid. Cell or organism with three sets of chromosomes.

USDA-ARS (United States Department of Agriculture - Agricultural Research Service). The principal in-house research agency of the United States Department of Agriculture (USDA). ARS is one of four agencies in USDA's Research, Education and Economics mission area. ARS is charged with extending the nation's scientific knowledge and solving agricultural problems.

Zygote. Product of the union or fusion of male and female gametes, often called a fertilized egg.